Walid Kaakeh

Agricultural Data Sovereignty: Empowering Farmers for the Future

Walid Kaakeh

Agricultural Data Sovereignty: Empowering Farmers for the Future

A Multidisciplinary Exploration of Ethics, Policy, and Industry in Digital Agriculture

Noor Publishing

Imprint

Cover image: www.ingimage.com

Publisher:
Noor Publishing
is a trademark of
Dodo Books Indian Ocean Ltd. and OmniScriptum S.R.L publishing group

120 High Road, East Finchley, London, N2 9ED, United Kingdom
Str. Armeneasca 28/1, office 1, Chisinau MD-2012, Republic of Moldova, Europe
Managing Directors: Ieva Konstantinova, Victoria Ursu
info@omniscriptum.com

Printed at: see last page
ISBN: 978-620-8-87064-5

Agricultural Data Sovereignty
Empowering Farmers for the Future of Food
A Multidisciplinary Exploration of Ethics, Policy, and Industry in Digital Agriculture

By
Prof. Dr. Walid Kaakeh
International Agricultural Consultant & Expert

First Edition – 2025

بسم الله الرحمن الرحيم

IN THE NAME OF ALLAH
THE MOST BLESSED, THE MOST MERCIFUL

Dedication

To my mother and father,
Aisha and Abdul Ghani Kaakeh
May God bless them and make them dwell
in His spacious gardens

To my wife
Fatima Khatib
for her unwavering patience, compassion, and most
importantly her love

To my beloved children,
Tasneem, Yasmine, Yamen, and Enas
who continue to be a source of inspiration and pride

I dedicate this book

Walid Kaakeh

Table of Contents

Preface

In the modern world, data has become as valuable as land, water, and labor in the agricultural sector. With the global population steadily increasing, the pressure on food systems has never been more pronounced. At the same time, the revolution in digital technologies and data analytics is reshaping how food is produced, distributed, and consumed. Precision agriculture, smart farming tools, satellite imaging, and big data analytics promise to enhance productivity, reduce waste, and make farming more sustainable. However, these advances also raise critical questions about ownership, control, and access to agricultural data. Who owns the data generated from the fields, from the satellites, from the sensors in the soil, and from the machinery used in modern farming practices? Can farmers retain control over this data, or will it be controlled by large corporations and technology companies?

The concept of agricultural data sovereignty has emerged as an urgent and necessary conversation at the intersection of ethics, policy, and industry practices. Data sovereignty refers to the right of farmers, communities, and nations to control their data, decide how it is used, and ensure that they benefit from the insights and innovations derived from it. The implications of this issue are vast, with profound impacts on food security, economic equity, and the future of global agriculture.

As technological advances rapidly transform the agricultural landscape, it is crucial that the voices of farmers—especially those from marginalized and smallholder communities—are heard in shaping the policies and frameworks that govern agricultural data. This book explores this very issue from a multidisciplinary perspective, bringing together academic theories, policy frameworks, and industry practices to address the pressing concerns around agricultural data sovereignty.

Throughout this book, the contributors explore the ethical questions surrounding the use of agricultural data, the legal and regulatory frameworks that are needed to safeguard farmers' rights, and the economic implications of data-driven agriculture. They examine the current state of agricultural data governance, the opportunities for inclusive growth, and the challenges that arise when powerful industry players control critical agricultural data. Moreover, this book looks to the future, considering how collaboration among various stakeholders—including farmers, technologists, policymakers, and civil society—can lead to a more equitable and sustainable agricultural data ecosystem.

As we move into the digital age, it is imperative that data does not become a tool of exploitation but rather an instrument for empowerment. By equipping farmers with control over their data, we can ensure that they are not only participants in the digital revolution but also beneficiaries of its potential. This book offers crucial insights for policy advocates, industry leaders, and academics alike, and provides a roadmap for navigating the complex ethical, legal, and economic terrain of agricultural data sovereignty.

The chapters that follow are a critical step in addressing this challenge, drawing attention to the need for a global framework that balances the potential of new technologies with the protection of farmers' rights and interests. Through a combination of theoretical perspectives, practical case studies, and forward-thinking solutions, this book seeks to chart a path forward in one of the most important issues facing the agricultural sector today.

It is my hope that this book will not only inform but inspire action toward a future in which agricultural data sovereignty is a reality for all farmers, ensuring that data-driven agriculture contributes to the prosperity, sustainability, and equity of global food systems.

Abstract

Agricultural data sovereignty has become a pivotal issue in the modern agricultural landscape, as new technologies transform farming practices and data plays an increasingly central role in decision-making. This book explores agricultural data sovereignty from three primary perspectives: academic, policy-oriented, and industry-focused. It addresses the ethical, legal, and economic challenges related to data ownership, privacy, and control in the agricultural sector, particularly concerning smallholder farmers. By examining the technological advancements that have enabled the data revolution in agriculture and the regulatory frameworks that govern data use, the book offers insights into how data sovereignty can be safeguarded. It also delves into the economic value of agricultural data and the risks posed by data monopolies, advocating for inclusive access to data and collaboration among stakeholders. With a focus on the future of agriculture, the book concludes with strategic recommendations to ensure that farmers' rights are protected, and that data-driven agriculture benefits all participants, fostering global food security, economic empowerment, and environmental sustainability.

Introduction

In recent decades, the agricultural industry has experienced a dramatic transformation driven by the digitization of farming practices. The advent of precision agriculture, smart farming technologies, and big data analytics has revolutionized the way crops are grown, managed, and harvested. However, this rapid technological development has raised a fundamental question: who owns and controls agricultural data? The issue of agricultural data sovereignty has emerged as a critical concern in shaping the future of farming, especially for smallholder farmers who often lack the power to control the vast amounts of data generated by their operations.

At its core, agricultural data sovereignty refers to the right of farmers and communities to own, control, and protect the data that is generated through agricultural practices. It encompasses a range of ethical, legal, and policy issues surrounding the collection, sharing, and use of data in the agricultural sector. As technology companies, agribusinesses, and policymakers seek to harness agricultural data for the purposes of innovation, efficiency, and food security, the interests of farmers must be safeguarded to ensure that they benefit from these technological advances.

This book addresses the challenges and opportunities posed by agricultural data sovereignty from multiple perspectives. The academic lens focuses on understanding the theoretical frameworks and scientific underpinnings that inform data sovereignty practices, while the policy-oriented perspective examines the legislative and regulatory approaches needed to ensure farmers' rights are protected. The industry-focused viewpoint explores the role of technology companies and agribusinesses in the data ecosystem and provides insights into how businesses can operate responsibly while respecting the rights of farmers.

In Chapter 1, we lay the foundation by introducing the concept of agricultural data sovereignty, defining its scope and highlighting the importance of farmers' rights in an increasingly data-driven agricultural landscape. We then examine the technological advancements that have enabled the data revolution in farming, exploring how smart agriculture, precision tools, and data analytics are reshaping the sector. The ethics of agricultural data use, particularly in terms of privacy, consent, and ownership, are critically discussed, establishing the ethical framework that underpins the book.

The policy landscape is explored in Chapters 2 to 4, where we delve into the legal and regulatory frameworks that influence agricultural data sovereignty. We explore the national and international policies that govern data collection, sharing, and protection, and assess how various jurisdictions approach the issue of data sovereignty. By examining both the global and local regulatory environments, we offer insights into how policymakers can ensure that data practices align with farmers' rights and equitable access to data benefits.

The economic implications of agricultural data are discussed in Chapters 5 and 6, focusing on the ways data is monetized and the economic value it generates. We also explore the potential risks posed by the monopolization of data by large agribusinesses and technology firms, analyzing how data ownership can shape the future of agricultural markets. The impact of data on smallholder farmers, who often lack the resources to engage with the digital economy, is critically examined, and solutions are proposed to promote inclusive access to agricultural data.

The importance of collaboration between different stakeholders in the agricultural data ecosystem is a recurring theme in the book. In Chapter 7, we explore the role of public-private partnerships in ensuring that agricultural data is managed responsibly and in a manner that benefits all stakeholders. Farmers' cooperatives and grassroots organizations are highlighted as key players in advocating for farmers' rights and providing support for data sovereignty efforts.

In Chapters 8 and 9, we examine the technological solutions and tools that can empower farmers to take control of their data. From blockchain to artificial intelligence, we explore how emerging technologies can facilitate data management, ensure transparency, and guarantee that farmers retain ownership over their data. Practical examples of successful implementation of these technologies in different regions are presented, showcasing their potential to foster a data-driven agricultural revolution.

The final chapters focus on the future of agricultural data sovereignty. In Chapters 10 to 12, we discuss the long-term implications of data sovereignty for global food security, environmental sustainability, and ethical farming practices. The role of education and capacity building is emphasized, with recommendations for fostering digital literacy and enhancing farmers' engagement with data. The conclusion presents a vision for the future, advocating for a global framework that balances technological innovation with the protection of farmers' rights and the promotion of sustainable agriculture.

Ultimately, this book seeks to provide a comprehensive examination of the issue of agricultural data sovereignty, offering insights and recommendations for policymakers, academics, and industry leaders. As the agricultural sector continues to evolve in the digital age, it is essential that data practices are grounded in ethical principles, that farmers retain control over their data, and that the benefits of data-driven agriculture are shared equitably across all stakeholders.

Author

Prof. Dr. Walid Kaakeh

Chapter 1: Defining Agricultural Data Sovereignty

1.1 Conceptualizing Data Sovereignty in Agriculture

- Provide a comprehensive definition: Agricultural data sovereignty refers to the right of individuals, communities, and nations to govern the collection, ownership, access, and use of agricultural data generated within their domains.
- Explore related terms such as data governance, data ownership, data stewardship, and digital rights.
- Contrast data sovereignty with data openness—discussing when open data is empowering versus when it can be exploitative.

1.2 Historical Context and Evolution of Data Sovereignty

- Trace the rise of agricultural digitization from the Green Revolution to the Fourth Industrial Revolution.
- Describe how traditional agricultural knowledge systems (e.g., indigenous knowledge) were often extracted without benefit-sharing—drawing a parallel to today's data practices.
- Introduce early warning signs in precision agriculture and agri-platforms that prompted the call for data justice and sovereignty.

1.3 Key Stakeholders in Agricultural Data Governance

- **Farmers**: The primary data producers—yet often least empowered in data ecosystems.
- **Agri-Tech Companies**: Innovators and platform owners who collect, analyze, and monetize farm data.
- **Governments**: Regulators and custodians of public interest—tasked with creating enabling legal frameworks.

- **Research Institutions**: Academics and scientists who use agricultural data for innovation and food security modeling.
- **Civil Society and Indigenous Groups**: Advocates for ethical data use, rights-based approaches, and inclusion.
- **Investors and International Donors**: Influence data infrastructures through financing and development agendas.

1.4 Categories of Agricultural Data

- **Farm-Level Data**: Soil quality, yield, crop health, machinery usage, labor.
- **Market and Supply Chain Data**: Pricing, demand forecasts, logistics.
- **Environmental and Geospatial Data**: Climate, satellite imagery, biodiversity.
- **Genetic and Breeding Data**: Crop and livestock genetics, biotechnological developments.

1.5 Frameworks for Understanding Sovereignty

- Political: National sovereignty over data versus global cloud infrastructure.
- Legal: Intellectual property, data protection laws, and rights-based data frameworks.
- Ethical: Principles of fairness, justice, informed consent, and benefit sharing.
- Technological: The role of open-source platforms, data standards, and privacy-enhancing technologies.

Chapter 2: The Ethics of Agricultural Data Use

Overview

This chapter explores the ethical dimensions of agricultural data management, focusing on principles such as autonomy, justice, transparency, accountability, privacy, and consent. It addresses how ethical lapses in the collection, handling, and sharing of data can affect farmers, agribusinesses, researchers, and rural communities.

2.1 Ethical Considerations in Data Collection, Ownership, and Sharing

- **Data Collection**: How is agricultural data collected? Who is collecting it, and for what purpose?
    - Are farmers aware of the data being collected by smart devices, sensors, or mobile apps?
    - Are there clear terms and conditions?
- **Ownership**:
    - Who owns the data—the farmer, the technology provider, the government, or a third-party service?
    - Is data treated as an individual right, a communal asset, or a public good?
- **Data Sharing**:
    - Should agricultural data be shared freely for research and development?
    - What boundaries should be set around commercial exploitation vs. public benefit?
- Introduce the concept of data stewardship—the idea that responsible parties manage data on behalf of data subjects with care and respect.

2.2 Privacy, Consent, and Transparency in Agricultural Data Practices

- **Informed Consent**:
  - Farmers often lack technical knowledge to fully understand what they are agreeing to when using agtech platforms or tools.
  - This section calls for plain language agreements, informed choices, and the right to opt-out or revoke consent.
- **Privacy Risks**:
  - Examples: Geospatial data revealing land value, or yield data showing production levels—these can be used by competitors or speculators.
  - Surveillance concerns: Continuous data collection without adequate safeguards can violate privacy.
- **Transparency**:
  - Transparency in how data is collected, stored, used, and monetized.
  - The need for auditable systems and data flow diagrams to ensure ethical clarity.

2.3 Implications for Farmers and Communities

- Farmers, especially in low-resource settings, are vulnerable to data misuse.
  - Example: An app collects pest data but also sells farmer input trends to agrochemical companies, influencing pricing.
- **Community Data**:
  - Local knowledge, indigenous data, and traditional farming practices are often digitized without consent or compensation.
  - There's a need for collective data rights and community data governance frameworks.

- **Equity Issues**:
    - Ethical concerns about how data systems may reproduce or deepen social inequalities (e.g., gender bias, lack of access for smallholder women farmers).
    - Highlight how digital tools must be inclusive and representative of diverse stakeholders.

2.4 Case Studies and Ethical Dilemmas

- Include real or hypothetical case studies to show how ethics plays out in practice:
    - A government deploys drones to collect agricultural data but doesn't inform farmers.
    - A private firm uses satellite imagery to predict harvests and speculate in commodity markets without local input.
    - An open-data platform inadvertently exposes sensitive data from indigenous lands.
- Provide **reflection questions** for academic, policy, and industry readers:
    - How do we balance innovation with rights?
    - What ethical frameworks can guide responsible data use?
    - How do we ensure informed participation from all stakeholders?

2.5 Toward an Ethical Agricultural Data Ecosystem

- This section proposes guiding ethical principles for the sector:
    - Respect for data subjects (farmers, communities).
    - Justice and fairness in data access and benefit-sharing.
    - Accountability mechanisms for misuse or abuse.
    - Sustainability—ensuring data governance supports long-term agricultural resilience.

- Discuss codes of ethics and standards that can be adopted across institutions (e.g., industry ethics charters, academic research guidelines, government protocols).

Chapter 3: Global Perspectives on Agricultural Data Sovereignty

Overview

This chapter broadens the lens to examine how different regions and countries conceptualize, implement, or struggle with agricultural data sovereignty. It contrasts developed and developing contexts, explores cultural and legal differences, and provides case studies that illustrate how sovereignty over agricultural data is handled across the globe. It also highlights how power, access, and inequality intersect with global data flows in agriculture.

3.1 Comparative Analysis: Global Data Governance Models

This section surveys how countries or regions govern agricultural data, including:

- **The European Union (EU)** and its General Data Protection Regulation (GDPR) and implications for farm data.
- **United States**: A largely market-driven approach with strong private-sector leadership and minimal federal oversight in agricultural data.
- **China**: Centralized data strategies linked to national security and agri-tech competitiveness.
- **African Union and Indigenous Perspectives**: Emphasis on communal rights, land-linked data ownership, and food sovereignty movements.

The focus is on how data protection laws, data localization policies, and open data initiatives shape the sovereignty landscape.

3.2 Case Studies from the Global South and Global North

Here, you can present comparative case studies such as:

- **Kenya**: Use of digital platforms for crop insurance and market access, and the challenges of ensuring farmer data protection.

- **India**: National digital agriculture missions and debates on private–public partnerships and farmer rights over data.
- **Brazil**: Tensions between agribusiness expansion and indigenous data sovereignty in Amazonian agriculture.
- **Germany or Netherlands**: Highly digitized farms with legal frameworks supporting farmers' control over data.

Each case can highlight lessons on:

- Inclusivity (who benefits from digital agriculture),
- Power asymmetries (who holds decision-making authority),
- Data colonialism (when data is extracted without equitable returns), and
- Community-led governance.

3.3 Lessons from Other Sectors: Health, Environment, and Technology

This part draws **parallels** between agricultural data and:

- **Health data**: How informed consent and ethical data use are standard in healthcare—and what agriculture can learn.
- **Environmental data**: Challenges in defining public vs. private ownership of shared environmental resources like water or biodiversity data.
- **Technology sector**: The rise of platform monopolies and surveillance capitalism in tech, mirrored now in digital agriculture.

It also explores how cross-sector collaboration could inform a rights-based **approach** to agricultural data sovereignty.

Key Takeaways

- Data sovereignty is context-specific, influenced by political, legal, economic, and cultural factors.
- Developing countries face greater risks of data dependency and digital colonization.
- There's a growing need for localized, culturally relevant frameworks that balance innovation with protection.

Chapter 4: The Role of Technology and Innovation

Overview

This chapter explores how emerging technologies such as artificial intelligence (AI), the Internet of Things (IoT), drones, satellite imagery, blockchain, and big data analytics are transforming agricultural practices. It highlights both the opportunities and risks these technologies pose in relation to data sovereignty and ethics.

4.1 The Influence of AI, Big Data, and IoT in Agriculture

- **AI and Machine Learning**: Algorithms analyze vast amounts of data for decision-making (e.g., crop predictions, pest detection, irrigation scheduling).
  - Who trains these algorithms?
  - Are the data inputs culturally or regionally biased?
  - Are the benefits equitably distributed?
- **Big Data Analytics**: Cloud-based platforms aggregate and analyze data from millions of farmers.
  - Raises questions of centralization, monopolization, and farmer autonomy.
- **IoT Devices and Sensors**: Soil sensors, weather stations, and livestock trackers collect real-time data.
  - Who has access to the data being generated continuously?
  - Are farmers being surveilled unknowingly?

4.2 Opportunities and Risks of Technological Advancements *(continued)*

Opportunities:

- **Improved productivity and sustainability**: Data-driven tools can optimize resource use (e.g., water, fertilizer), boost yields, and reduce environmental footprints.
- **Climate resilience**: Tech-enabled early warning systems, weather forecasting, and risk mapping support farmers in adapting to climate change.
- **Market access and financial inclusion**: Digital platforms can link farmers to markets, provide credit scoring using data, and facilitate crop insurance.
- **Knowledge democratization**: Mobile applications and open platforms can spread agronomic knowledge more widely and inclusively—if designed with equity in mind.

Risks:

- **Technological dependency**: Overreliance on proprietary tools may disempower farmers and erode traditional knowledge.
- **Data monopolies**: Large corporations may dominate platforms and control entire agri-data ecosystems.
- **Exclusion and marginalization**: Smallholders and Indigenous farmers may be excluded from digital agriculture due to lack of access, literacy, or infrastructure.
- **Algorithmic bias**: Data-driven tools may inadvertently reinforce inequalities if they reflect biased datasets or fail to consider local contexts.

4.3 Data Security, Protection, and Ethical Tech Development

This section focuses on the need for secure, transparent, and ethical development of agri-tech systems.

- **Cybersecurity risks**: Threats from hacking, ransomware, or data breaches can compromise sensitive farm data and infrastructure.
- **Ethical design principles**:
  - Embedding privacy-by-design and ethics-by-design approaches.

    - Creating user-centric systems that prioritize informed consent, control, and opt-out mechanisms for farmers.
- **Open-source vs. proprietary systems**: The balance between innovation and openness; how open-source tools can empower communities and reduce dependency on corporate platforms.
- **Blockchain in data sovereignty**: The potential role of blockchain in enabling transparent, tamper-proof, farmer-controlled data systems.

Data Security

- Discuss the importance of robust cybersecurity protocols to protect sensitive farm data from breaches, leaks, or misuse.
- Explore risks like:
    - Unauthorized access to data stored in cloud platforms.
    - Hacking of IoT-connected equipment (e.g., remote manipulation of irrigation systems or drones).
    - Ransomware attacks targeting agricultural data systems.

Data Protection Frameworks

- Evaluate whether current legal and technical safeguards are adequate.
- Recommend implementing standards like encryption, access control, data anonymization, and audit trails.
- Examine how Data Protection Impact Assessments (DPIAs) can be applied in agricultural tech deployment.

Ethical Technology Development

- Promote inclusive, participatory design of technologies that involve farmers—especially women, smallholders, and indigenous groups—early in the design and testing process.
- Emphasize the concept of "ethical by design": ensuring that technologies are built with ethics, privacy, and data ownership in mind from the start—not as an afterthought.

- Advocate for open-source and farmer-owned platforms as more democratic alternatives to proprietary, corporate systems.

Summary

Technological innovation is transforming agriculture into a highly data-driven enterprise. From sensors in the field to algorithms in the cloud, technology offers tools for smarter, more efficient farming—but it also introduces complex ethical, legal, and social implications.

- While AI, big data, and IoT promise gains in yield, climate adaptation, and market access, they can also centralize control, undermine farmer autonomy, and exacerbate inequality if not guided by ethical and inclusive frameworks.
- Issues such as data ownership, bias in algorithms, exclusion of marginalized groups, and digital surveillance demand strong data protection laws, community-based innovation, and secure tech systems.
- The chapter concludes with a call for ethical-by-design technology and farmer-centered digital ecosystems—to ensure that innovation serves the people who grow the food.

Chapter 5: Legal and Policy Frameworks Supporting Data Sovereignty

Overview

This chapter examines how legal instruments, regulatory frameworks, and policy strategies at local, national, and international levels are evolving to support (or hinder) agricultural data sovereignty. It analyzes current laws, identifies gaps, and offers recommendations for developing policies that empower farmers and rural communities to retain control over their data while fostering innovation, collaboration, and accountability.

5.1 Existing Legal Frameworks: Data Protection, Privacy, and Sovereignty Laws

- **General Data Protection Laws**:
    - Examples: GDPR (EU), CCPA (California), Kenya's Data Protection Act, etc.
    - How these laws apply (or don't) to agricultural data, which often falls into grey areas between personal and non-personal data.
    - Limitations in addressing group rights, communal data, and non-human-centric data (like soil or satellite data).
- **Sovereignty-Driven Policies:**
    - Data localization mandates: Requiring data to be stored within national borders—intentions, controversies, and implications for global agribusinesses.
    - Policies addressing food sovereignty, land rights, and digital sovereignty, especially in Global South contexts.
- **Sector-Specific Gaps:**
    - Lack of explicit regulation or recognition of farm-level data rights.
    - Absence of legal clarity on ownership vs. access vs. control.

5.2 Gaps and Challenges in the Legal Landscape

- **Ambiguity of ownership**: Legal systems often fail to define who owns agricultural data—the farmer, the equipment provider, the platform host?
- **Multistakeholder complexity**: Disputes may arise between farmers, agritech companies, governments, and researchers over who can access and use what data.
- **Lack of enforcement mechanisms**: Even when legal protections exist, weak enforcement often leaves smallholders vulnerable.
- **Cross-border data flows**: Legal frameworks struggle to govern international data transfers, especially when cloud services are global.

5.3 Policy Innovations and Best Practices

- **Open data charters**: Policies that define and promote ethical open data sharing (e.g., the GODAN initiative).
- **Model legal instruments**: Development of model "farm data contracts" that define roles, rights, and responsibilities.
- **National digital agriculture strategies**:
  - Case examples: India's Digital Agriculture Mission, Brazil's Agro 4.0, Ethiopia's National Digital Strategy.
  - Analysis of how these integrate data sovereignty principles (or fail to).
- **Decentralized governance**:
  - Encouraging community-level or farmer-led data governance models.
  - Examples of data trusts, data cooperatives, and platform co-ops in agriculture.

5.4 Toward Inclusive, Ethical Policy-Making

- **Stakeholder engagement**: Emphasizing bottom-up, participatory policy development, especially involving farmers, Indigenous groups, women, and youth.
- **Cross-sector collaboration**: Aligning laws and policies across agriculture, data science, environment, and human rights sectors.
- **Ethical and equitable frameworks**: Moving beyond compliance toward justice-oriented policy—where farmers are not just data subjects, but active decision-makers.
- **Interoperability with global frameworks**: Aligning national laws with international standards while respecting local customs and sovereignty.

Summary

Laws and policies play a foundational role in shaping who benefits from agricultural data. However, most existing frameworks lag behind technological innovation and fail to protect the rights of those who generate data—primarily farmers and rural communities.

This chapter calls for a shift toward:

- Clear, enforceable legal rights around data ownership, access, and use.
- Inclusive policy-making that respects Indigenous knowledge systems and communal data models.
- International cooperation that balances innovation with equity and sovereignty.

Chapter 6: Ethical Dimensions of Agricultural Data Governance

Overview

This chapter explores the ethical principles and dilemmas that arise in the collection, ownership, sharing, and use of agricultural data. It addresses the moral responsibilities of stakeholders—including governments, researchers, agribusinesses, and technology providers—toward farmers, rural communities, and ecosystems. It also examines emerging ethical frameworks and value-based approaches that aim to guide fair, inclusive, and responsible agricultural data practices.

6.1 Core Ethical Principles in Agricultural Data Governance

- **Autonomy and Consent**:
    - Farmers' right to make informed decisions about whether and how their data is collected, used, or shared.
    - Need for informed consent protocols, written in accessible language, especially for smallholders and Indigenous communities.
- **Justice and Equity**:
    - Ensuring fair distribution of the benefits derived from data use.
    - Addressing power imbalances between large agri-tech firms and smallholder farmers.
- **Transparency and Accountability**:
    - Clear information on who collects, who owns, and who profits from data.
    - Establishing transparent data governance frameworks and dispute resolution mechanisms.
- **Respect for Cultural and Local Knowledge**:

    - Recognizing the value of traditional farming practices and community-held knowledge.
    - Avoiding extractive data practices that treat local knowledge as a commodity without fair compensation.

6.2 Ethical Dilemmas in Agricultural Data Use

- **Consent vs. Coercion**:
    - Many digital platforms require data sharing as a condition for access, effectively limiting genuine consent.
- **Privacy vs. Public Good**:
    - Balancing the need to protect individual or community privacy with the value of shared data for research, innovation, and food security.
- **Ownership vs. Openness**:
    - Tensions between protecting farmer-owned data and promoting open data for public benefit (e.g., climate models, pest outbreak prediction).
- **Profit vs. Participation**:
    - How profits generated from agri-data often benefit private companies more than the farmers who produce the data.

6.3 Emerging Ethical Frameworks and Guidelines

- **FAO's Code of Conduct on Data for Agriculture**: Principles for responsible digital agriculture, promoting inclusiveness, transparency, and accountability.
- **Principles for Digital Development**: Including design with the user, address privacy and security, and be collaborative.
- **CARE Principles for Indigenous Data Governance**:
    - Collective Benefit, Authority to Control, Responsibility, and Ethics—providing a strong foundation for Indigenous data rights.

- **Rights-Based Approaches**:
  - Anchoring data governance in human rights, land rights, and economic justice.
  - Moving beyond corporate social responsibility (CSR) toward enforceable ethical commitments and community-led governance.

Summary

This chapter reinforces the importance of embedding ethics at the core of agricultural data governance. Ethical principles like consent, equity, transparency, and justice are not just ideals—they must be operationalized through policy, design, and daily practices. Recognizing the inherent power dynamics in agri-data systems and working to correct them is essential for a just and sustainable agricultural future.

Chapter 7: Stakeholder Roles and Responsibilities in Data Ecosystems

Overview

This chapter delves into the diverse array of stakeholders involved in the agricultural data ecosystem. It outlines their roles, responsibilities, and the ethical considerations each group must account for in ensuring the fair, secure, and sustainable use of agricultural data. Stakeholders discussed in this chapter include farmers, agri-tech companies, governments, research institutions, and non-governmental organizations (NGOs), among others. It also explores the need for collaboration and multi-stakeholder engagement in data governance.

7.1 The Role of Farmers and Rural Communities

- **Data Providers**: Farmers are often the primary producers of agricultural data, whether through direct measurement (e.g., soil moisture, weather data) or through their use of digital tools like GPS systems, sensors, or crop management software.
- **Data Ownership**: The right of farmers to control their data and decide who has access, how it is used, and for what purposes. Emphasis is placed on the ethical obligation to respect local sovereignty and community knowledge.
- **Capacity Building**: Farmers need ongoing support and training to understand and use data systems effectively. Providing them with tools to manage, interpret, and protect their own data is crucial for empowering them in the digital age.

7.2 The Role of Agribusinesses and Agri-Tech Companies

- **Data Collectors and Processors**: Agri-tech companies are often responsible for collecting, processing, and monetizing agricultural data.

Their role is essential in facilitating innovation through the use of AI, machine learning, and predictive analytics, but it also carries ethical obligations regarding data privacy, security, and fair profit-sharing.

- **Platform Providers**: Many agri-tech companies offer platforms where farmers can access digital tools. These platforms should be designed with farmer-centric principles in mind, ensuring interoperability, data privacy, and transparency in data usage.
- **Business Ethics**: The role of agri-tech companies in ensuring their business models align with sustainability, equity, and fair data-sharing practices. They must avoid exploiting data from farmers for commercial **gain** without providing fair compensation or access to the benefits derived from the data.

7.3 The Role of Governments and Policymakers

- **Regulators and Enforcers**: Governments have the responsibility to create and enforce policies that safeguard farmers' data rights, promote data sovereignty, and regulate the ethical use of agricultural data. This includes crafting data protection laws (such as GDPR, CCPA) and food sovereignty policies.
- **Enablers of Collaboration**: Governments should also play a facilitating role in fostering **collaboration** between farmers, businesses, researchers, and other stakeholders, ensuring that policies foster equitable partnerships.
- **Support for Data Infrastructure**: Investment in national and regional infrastructure for data collection, storage, and analytics is necessary for both public and private entities to share valuable insights.
- **Public Good vs. Private Interests**: Striking a balance between supporting **economic growth and promoting public good through open access to** agricultural data while respecting privacy and local knowledge.

7.4 The Role of Research Institutions and Academia

- **Knowledge Providers**: Research institutions are integral in generating data, insights, and methodologies for agricultural innovations. They can bridge the gap between scientific research and practical application in the field.
- **Open Data and Collaboration**: Universities and research organizations should lead efforts to share agricultural data openly and collaborate with governments, agribusinesses, and farmers to maximize social benefits from data-driven discoveries.
- **Ethical Research**: Research must adhere to ethical standards regarding consent, privacy, and equitable participation. Researchers should also ensure that local communities and smallholder farmers have a voice in research design and dissemination.

7.5 The Role of Non-Governmental Organizations (NGOs) and Civil Society

- **Advocacy for Farmers' Rights**: NGOs are crucial in advocating for the rights of farmers—particularly smallholder, marginalized, and Indigenous communities. They can play a pivotal role in ensuring that data sovereignty is not undermined by large corporations or powerful governments.
- **Capacity Building and Education**: NGOs are instrumental in raising awareness among farmers about their rights regarding data use, including informing them about data ownership and ethical data-sharing practices.
- **Community Data Governance**: NGOs can support community-based data governance models and advocate for local data sovereignty initiatives, ensuring that farmers' voices are central in any policy or technological solutions.

7.6 The Role of International Organizations and Global Initiatives

- **International Guidelines and Standards**: International organizations like the FAO, UNDP, and World Bank play key roles in promoting global data standards and creating frameworks for agricultural data governance. These guidelines can help create a global consensus on data privacy, data sharing, and data sovereignty.
- **Global Data Sharing Initiatives**: Global data initiatives, such as the Global Open Data for Agriculture and Nutrition (GODAN) initiative, encourage the open sharing of agricultural data for the collective good while ensuring ethical guidelines are followed.
- **International Policy Harmonization**: International entities help harmonize policies related to agricultural data governance to ensure consistency across borders, especially in a world of cross-border data flows.

Summary

Chapter 7 underscores the interconnectedness of the stakeholders within the agricultural data ecosystem. Each group has a critical role in ensuring that data sovereignty, ethical data practices, and equitable outcomes are realized. While the power dynamics between farmers, tech companies, governments, and research institutions differ, all must work together toward the goal of a fair and transparent agricultural data landscape.

Chapter 8: Emerging Technologies and Future Trends in Agricultural Data

Overview

In this chapter, we explore the cutting-edge technologies transforming the agricultural data landscape, from artificial intelligence (AI) and blockchain to Internet of Things (IoT) devices and precision agriculture tools. These advancements are reshaping the way data is collected, processed, and utilized, enabling greater efficiency, sustainability, and resilience in agriculture. The chapter also examines future trends in agricultural data governance, focusing on interoperability, data-driven decision-making, and the evolving global data ecosystem.

8.1 Artificial Intelligence and Machine Learning in Agriculture

- **Predictive Analytics**: AI is revolutionizing agriculture through the ability to predict weather patterns, crop yields, and even pest infestations. By analyzing historical data and real-time inputs, AI systems can provide farmers with timely insights to optimize their operations.
- **Automation and Decision Support**: AI-enabled platforms now assist farmers in decision-making by automating tasks such as crop monitoring, irrigation management, and pest control, leading to more efficient use of resources and higher yields.
- **Data-Driven Solutions**: Machine learning algorithms process vast amounts of agricultural data to identify patterns and correlations that can inform everything from crop breeding to resource management.

8.2 Blockchain Technology for Data Security and Transparency

- **Data Integrity and Trust**: Blockchain's distributed ledger technology is enhancing the security, transparency, and traceability of agricultural data.

This can address concerns around data privacy and ownership by providing immutable records of data transactions.

- **Supply Chain Transparency**: Blockchain ensures that data related to food production (e.g., sourcing, quality control, and distribution) is secure and transparent, building trust between farmers, consumers, and retailers.
- **Smart Contracts**: Blockchain technology can enable smart contracts for agriculture, automating transactions, agreements, and payments, thereby reducing administrative costs and improving efficiency in data exchanges.

8.3 Internet of Things (IoT) in Precision Agriculture

- **Sensor Networks**: IoT-enabled sensors deployed across fields can provide farmers with real-time data on variables such as soil moisture, temperature, pH levels, and nutrient content. This continuous flow of data helps in precise decision-making for irrigation, fertilization, and pest management.
- **Remote Monitoring**: IoT devices, combined with satellite imagery and drones, allow for remote monitoring of crops and livestock, enabling farmers to track health conditions and environmental stresses from afar.
- **Automation and Efficiency**: IoT devices, like autonomous tractors and drones, are enabling automated farming processes, reducing labor costs and increasing efficiency. These tools are designed to collect vast amounts of operational data, feeding into agricultural management systems for real-time decision support.

8.4 Big Data and Cloud Computing in Agriculture

- **Data Aggregation and Analysis**: The ability to aggregate and analyze massive datasets through cloud computing is empowering farmers with insights that were previously out of reach. Big data platforms integrate weather data, soil data, market trends, and satellite imagery to provide holistic views of farm operations.

- **Predictive and Prescriptive Analytics**: With the increasing availability of data, farmers can use cloud-based systems to predict future trends and make prescriptive recommendations. This helps to mitigate risks and improve supply chain management.
- **Collaborative Platforms**: Cloud computing enables farmers, researchers, and agribusinesses to collaborate and share agricultural data and insights, improving collective decision-making and fostering innovation.

8.5 Remote Sensing and Satellite Technologies

- **Precision Crop Monitoring**: Satellite imagery, along with remote sensing technologies, allows for detailed analysis of crop health, water stress, and growth patterns across large agricultural areas. This technology is particularly beneficial for precision farming, enabling targeted interventions in real time.
- **Geospatial Data for Landscape Management**: Remote sensing provides geospatial data that can be used to manage landscapes effectively, helping farmers optimize field boundaries, plan crop rotation, and minimize environmental impact.
- **Early Detection of Pests and Diseases**: Remote sensing and satellites can detect early warning signs of pest infestations or disease outbreaks by analyzing anomalies in plant health, enabling rapid response to minimize damage.

8.6 Future Trends in Agricultural Data Governance

- **Interoperability and Data Integration**: As technology evolves, there will be increasing demand for interoperability between different platforms and data sources. This includes integrating data from IoT sensors, cloud platforms, satellites, and AI systems into cohesive, accessible formats that can be easily shared among stakeholders.

- **Data Sovereignty and Ethics**: The increasing use of data from diverse stakeholders raises significant concerns around data sovereignty. Future trends point to stronger national and international frameworks that ensure ethical practices, data protection, and equitable benefits for all participants, particularly farmers in developing regions.
- **Blockchain and Data Ownership**: As data becomes a critical asset in agriculture, blockchain and other decentralized technologies are likely to play a bigger role in ensuring that farmers retain ownership of their data and can monetize it in ethical ways.
- **Artificial Intelligence and Automation**: AI will continue to advance, enabling even more automated farming systems that reduce human intervention. This will open up new possibilities in precision farming, resource efficiency, and climate resilience, but will also require robust policies on data ethics and equity.

Summary

Chapter 8 introduces the transformative impact of emerging technologies like AI, blockchain, IoT, and remote sensing in the agricultural data ecosystem. These technologies not only enhance efficiency and productivity but also raise important questions around data governance, ownership, and ethics. As we look to the future, it's clear that collaboration, interoperability, and data sovereignty will be critical themes in shaping a sustainable, ethical, and equitable agricultural data landscape.

Chapter 9: Policy and Regulatory Frameworks for Agricultural Data Sovereignty

Overview

This chapter explores the policy and regulatory frameworks that govern agricultural data sovereignty, focusing on the evolving role of governments, international organizations, and industry leaders in shaping the legal and ethical landscape of agricultural data. We'll examine how regulations can ensure data ownership, privacy, and access, while balancing the interests of farmers, businesses, and global stakeholders. Furthermore, it will highlight key policy recommendations aimed at promoting equitable data practices, security, and sustainable development in the agricultural sector.

9.1 The Concept of Agricultural Data Sovereignty

- **Definition and Significance**: Agricultural data sovereignty refers to the right of individuals and communities—especially farmers and indigenous groups—to control, manage, and protect their agricultural data. This principle is essential in ensuring that agricultural data is not exploited by external actors without consent, and that it is used for the benefit of local communities.
- **The Global Context**: As agricultural data becomes increasingly valuable in global trade, governance, and technological innovation, the question of data sovereignty becomes more pressing. Countries are grappling with the challenge of balancing national sovereignty with the global flow of data.
- **Ethical Data Ownership**: At its core, agricultural data sovereignty is about ensuring that farmers and local communities are recognized as the true owners of their data, with the freedom to share, protect, or sell that data as they see fit.

9.2 International Policy Frameworks and Initiatives

- **The Role of the United Nations**: The FAO and other UN agencies are actively involved in developing international frameworks for the ethical governance of agricultural data. This includes efforts to create open data standards, encourage collaborative platforms, and foster data-sharing agreements that benefit smallholder farmers.
- **Global Data Sharing Initiatives**: Programs like Global Open Data for Agriculture and Nutrition (GODAN) aim to increase access to agricultural data globally while ensuring that such initiatives are conducted in a manner that respects data sovereignty. International organizations are working to balance open access to data with ethical concerns around data privacy and exploitation.
- **The Role of International Trade Agreements**: International trade agreements, such as those within the World Trade Organization (WTO), may influence how agricultural data is regulated across borders. These agreements need to integrate provisions on data privacy, security, and data sovereignty to avoid exploitation by multinational companies.

9.3 National and Regional Regulatory Approaches

- **Data Protection Laws**: Many countries have enacted or are in the process of enacting data protection laws that directly affect agricultural data. These laws ensure that personal and sensitive data related to farming operations is protected from misuse. Examples include the GDPR in the EU and CCPA in California, which offer frameworks for managing data consent and privacy.
- **Agricultural Data as Public Good**: Some countries, particularly in the Global South, are focusing on how agricultural data can be treated as a public good—not just as a commercial asset. National regulations are being

crafted to ensure that data sharing for research, climate change mitigation, and food security is prioritized over corporate interests.

- **Regulating Private Data Use**: Governments must enact laws that regulate how private companies collect and use agricultural data. These regulations should guarantee that farmers' data is not exploited for profit without fair compensation and that their rights to data access and ownership are upheld.

9.4 Balancing Private Interests with Public Good

- **Corporate Responsibility in Data Governance**: Multinational agribusinesses and tech companies must be held accountable for how they collect, use, and distribute agricultural data. While private companies drive technological innovation, there is an increasing push to align corporate actions with public good—ensuring that farmers and local communities benefit from technological advancements.
- **Data Profit-Sharing Models**: A growing number of policy discussions center around data profit-sharing—ensuring that farmers receive fair compensation for their role in generating agricultural data. Governments and international organizations may need to mandate shared benefits in data agreements, encouraging ethical data use in the commercial sector.
- **Open Access vs. Closed-Data Models**: One key challenge in agricultural data governance is determining the appropriate balance between open access to data for public good and data protection for individual farmers. Open data models benefit global food security, while closed data models can protect local interests. This balance is crucial to establishing ethical data policies.

9.5 National and International Collaboration on Data Sovereignty

- **Public-Private Partnerships**: Governments should collaborate with agri-tech companies, NGOs, and research institutions to develop inclusive and collaborative data frameworks. By working together, stakeholders can ensure data sovereignty and ethical data-sharing practices.

- **Multi-Stakeholder Dialogues**: Involving farmers, consumers, tech companies, and policymakers in multi-stakeholder dialogues is essential for creating transparent and inclusive data policies. These conversations should focus on fair access, security, and the benefits of data for smallholder farmers.
- **Global Standards for Agricultural Data Governance**: Efforts to create global standards for agricultural data are essential in creating consistency across borders. Standards should address concerns related to data privacy, ownership, security, and use, enabling farmers worldwide to share data without fear of exploitation.

9.6 Key Policy Recommendations

- **Establish Clear Data Ownership Rights**: Laws must be developed to ensure that farmers retain ownership of their agricultural data and are empowered to control how it is shared or monetized.
- **Promote Open Data for Research**: Governments and institutions should promote open data sharing for research purposes, especially in the areas of food security, climate change, and sustainable agriculture.
- **Support Farmers with Digital Literacy**: Policy initiatives should focus on empowering farmers with the tools and knowledge to manage their data, understand data rights, and benefit from emerging technologies.
- **Strengthen International Cooperation**: International policies should facilitate cross-border collaboration in agricultural data governance, allowing countries to share data while respecting national data sovereignty principles.

- **Encourage Ethical Data Profit-Sharing**: Governments and industry stakeholders should establish guidelines for data profit-sharing models, ensuring farmers are compensated for their data contributions.

Summary

Chapter 9 provides a comprehensive exploration of the policy and regulatory frameworks governing agricultural data sovereignty. From international guidelines and national laws to corporate responsibility, this chapter underscores the critical need for ethical, inclusive, and transparent data governance. It also highlights the importance of multi-stakeholder collaboration in fostering an environment where agricultural data can be used responsibly and fairly, ultimately benefiting all players in the data ecosystem.

Chapter 10: Building Ethical Data Ecosystems for Sustainable Agriculture

Overview

This chapter delves into the creation of ethical data ecosystems within the agricultural sector, emphasizing the importance of ensuring that agricultural data is used responsibly, sustainably, and with respect for data sovereignty. By focusing on both technological advancements and ethical frameworks, the chapter explores how these ecosystems can promote sustainability, innovation, and collaboration among various stakeholders. It also addresses the challenges that arise in creating such ecosystems and offers recommendations for building a data ecosystem that balances the needs of farmers, consumers, and the environment.

10.1 Understanding Ethical Data Ecosystems

- **Definition of Ethical Data Ecosystem**: An ethical data ecosystem is one where agricultural data is managed, shared, and utilized in ways that prioritize transparency, privacy, security, and fairness for all involved parties, especially farmers. This ecosystem involves collaboration among farmers, technology providers, government bodies, researchers, and consumers, ensuring that data use aligns with sustainable agricultural practices and supports data sovereignty.
- **The Ethical Dimensions of Data Use**: Ethical considerations involve ensuring that agricultural data is not used for exploitation or harm, but instead contributes to positive outcomes such as increased food security, better environmental management, and economic empowerment for farmers. Ethical frameworks guide how data is collected, shared, stored, and monetized.

10.2 The Role of Technology in Shaping Ethical Data Ecosystems

- **Advancements in Agri-Tech**: Emerging technologies such as precision agriculture, artificial intelligence (AI), and big data analytics are transforming the agricultural industry. These technologies offer insights and predictive models that can increase crop yields, improve resource management, and reduce environmental impact. However, their implementation must be accompanied by ethical guidelines to ensure that they benefit farmers, especially smallholders, and do not exacerbate inequalities.
- **Blockchain for Transparency and Security**: Blockchain technology has the potential to create transparent, secure, and tamper-proof agricultural data systems. It can ensure that farmers retain ownership of their data, that transactions are fairly conducted, and that data privacy is respected. By using blockchain, data sharing can be more transparent, and the traceability of agricultural products from farm to table can help build trust among consumers.
- **IoT and Sensor Technologies**: The use of Internet of Things (IoT) devices and sensors allows farmers to monitor and manage their farms in real-time, generating massive amounts of data. The ethical use of this data requires clear guidelines on data ownership, usage, and security, as well as mechanisms to ensure that farmers have control over how their data is used by third parties.

10.3 Collaborative Models for Data Sharing

- **Farmer-Centered Data Governance**: One of the core tenets of an ethical data ecosystem is ensuring that farmers are at the center of the data governance model. This involves not only informed consent for data collection but also providing farmers with the tools and education needed to understand how their data is being used. Farmers should be empowered to decide what data they wish to share and with whom, ensuring that they retain control over their information.

- **Cooperative Models**: **Cooperatives** or **farmers' organizations** can play a pivotal role in collectively managing agricultural data. By aggregating data from multiple farmers, cooperatives can negotiate better terms for data sharing and usage with tech companies, research institutions, and policymakers. They can also ensure that the value derived from the data is distributed equitably among the members.
- **Public-Private Partnerships**: Collaboration between governments, private companies, and research institutions is essential for building a data ecosystem that benefits all stakeholders. These partnerships can create open data platforms, ensure data privacy, and provide access to cutting-edge technologies for farmers, especially those in developing regions.

10.4 Ethical Challenges in Agricultural Data Ecosystems

- **Data Privacy and Security**: As more data is collected from farms, privacy and security concerns become more pronounced. Ensuring that sensitive data, such as personal and financial information, is protected is paramount. Technologies like encryption and access control measures must be implemented to protect farmers from potential breaches.
- **Data Monopolization**: There is a growing concern that large agribusinesses and tech giants could dominate the agricultural data ecosystem, thereby monopolizing access to vital agricultural information. This could lead to data exploitation without fair compensation for farmers. Regulatory frameworks must be designed to prevent such monopolies and ensure that data ownership remains decentralized.
- **Digital Divide and Inclusivity**: The digital divide between developed and developing regions, as well as between large agribusinesses and smallholder farmers, poses a significant challenge to building an ethical

data ecosystem. Access to technology, digital literacy, and data ownership must be addressed to avoid exacerbating inequalities.

10.5 Key Components of an Ethical Agricultural Data Ecosystem

- **Clear Data Ownership and Rights**: Ensuring that farmers own and control their data is a foundational aspect of an ethical data ecosystem. This includes setting clear rules around data access, usage, and monetization.
- **Data Transparency and Accountability**: Transparency in how agricultural data is collected, used, and shared is critical for building trust in the system. Accountability mechanisms should be in place to ensure that data use aligns with ethical guidelines and regulations.
- **Fair Data Profit-Sharing Models**: Ethical ecosystems must include models where farmers benefit financially from the use of their data. This can include profit-sharing agreements, where farmers receive compensation for allowing access to their data or for using data-driven services.
- **Environmental and Social Responsibility**: Ethical data ecosystems should align with sustainable agricultural practices and contribute to solving environmental issues. For instance, data-driven insights can help optimize water use, reduce pesticide dependence, and increase crop resilience, while also fostering social inclusivity by ensuring smallholder farmers have access to these tools.

10.6 Recommendations for Building Ethical Data Ecosystems

- **Implement Transparent Data Governance Policies**: Governments should establish clear, transparent policies for the management and governance of agricultural data, ensuring that the principles of data sovereignty and fairness are upheld.

- **Promote Collaborative Data Models**: Encourage the formation of cooperatives, alliances, and partnerships among stakeholders, ensuring that farmers' interests are protected while enabling the sharing of data for innovation and research.
- **Encourage Industry Standards for Data Sharing**: Industry leaders should collaborate to develop common standards for agricultural data, ensuring that data is shared in a way that respects privacy, security, and fair access.
- **Address the Digital Divide**: Governments and the private sector must work together to ensure that all farmers, regardless of their size or location, have access to digital tools and education on data rights and privacy.
- **Strengthen Legal Frameworks for Data Ownership**: Develop international agreements that reinforce data sovereignty and provide clear legal protections for farmers' ownership of their data.

Summary

Chapter 10 emphasizes the importance of creating ethical data ecosystems in agriculture that prioritize farmers' rights, privacy, data sovereignty, and sustainability. By focusing on technological innovations, collaborative models, and ethical governance, it lays out a vision for a future where agricultural data is used responsibly and to the benefit of both people and the planet. Key recommendations include promoting collaboration, developing data ownership frameworks, and addressing ethical challenges to ensure that the data ecosystem is fair and inclusive for all stakeholders.

Chapter 11: The Future of Agricultural Data Sovereignty: Emerging Trends and Challenges

Overview

Chapter 11 looks forward into the future of agricultural data sovereignty, examining the evolving landscape of data in agriculture, the emerging trends that will shape this future, and the challenges that stakeholders must navigate. As technological advances continue to reshape the way agricultural data is collected, managed, and used, it is essential to stay ahead of the curve to address the ethical, regulatory, and operational challenges that could arise. This chapter outlines potential scenarios, trends, and disruptions that could redefine agricultural data sovereignty and offers recommendations for addressing these challenges.

11.1 The Rapid Evolution of Agricultural Data Technologies

- **AI and Big Data Analytics**: The ongoing integration of artificial intelligence (AI) and big data into agriculture is revolutionizing how farms operate. Machine learning algorithms can predict crop performance, optimize planting schedules, and detect diseases early, which requires vast amounts of data. The ethical implications of these technologies, especially concerning data ownership and sovereignty, will only grow as they become more sophisticated.
- **IoT and Autonomous Systems**: As the Internet of Things (IoT) expands, the volume and variety of data collected from agricultural operations will increase exponentially. Autonomous systems, including drones and robots, will play a larger role in collecting data and making decisions on the farm. The question of who owns this data and how it will be used is critical, especially as AI-powered systems become autonomous in data analysis.
- **Precision Agriculture**: Precision agriculture, relying on technologies such as satellite imagery, remote sensing, and soil sensors, allows for highly

granular data collection on farm conditions. While these technologies offer great potential for improving farm productivity and environmental sustainability, they raise important issues regarding data privacy, accessibility, and security.

11.2 Global and Local Data Regulations: A Complex Landscape

- **International Data Sovereignty**: As agricultural data becomes a key resource in global trade and innovation, the issue of data sovereignty takes on an international dimension. Different countries have varying regulations governing data collection, storage, and cross-border transfer, creating a fragmented and complex regulatory landscape. This can lead to challenges for companies and policymakers working across borders, especially when handling sensitive farmers' data.
- **National Data Sovereignty**: National governments must ensure that their agricultural data regulations protect local farmers while also encouraging innovation. Some nations have begun to create their own data sovereignty frameworks, giving local farmers greater control over their data. These frameworks are still evolving and will require continual updates to adapt to the fast pace of technological change.
- **Data Localization Policies**: Data localization—the practice of requiring data to be stored and processed within the country where it was generated—may become more prevalent as governments seek to assert control over agricultural data. While these policies can strengthen **data sovereignty**, they may also limit the ability of companies to share data globally, creating a tension between national security and the global flow of information.

11.3 Emerging Trends in Data Ownership and Control

- **Farmer-Driven Data Models**: A significant emerging trend is the movement towards farmer-driven data governance, where farmers take the lead in managing their own data. This trend is supported by the increasing availability of decentralized technologies such as blockchain, which can provide secure, transparent, and immutable data storage systems that allow farmers to control access to their data and share it with trusted parties.
- **Cooperative Data Platforms**: **Data cooperatives** are likely to emerge as an effective model for smallholder farmers to collectively manage their data and protect their rights. By pooling resources and sharing data, farmers can ensure they receive equitable benefits and have a stronger negotiating position with data-driven companies and agricultural tech firms.
- **Smart Contracts and Data Monetization**: The rise of smart contracts, enabled by blockchain technology, will facilitate automated and transparent agreements between farmers, data consumers, and other stakeholders. This could enable farmers to directly monetize their data by creating digital assets that can be bought or sold within the ecosystem.

11.4 The Role of Public Policy in Shaping the Future of Data Sovereignty

- **Regulatory Innovation**: As the agricultural data landscape evolves, there is a need for innovative public policies that keep pace with technological change. This could involve creating frameworks for data ownership, fair use, and monetary compensation for farmers. Governments will need to be proactive in addressing the ethics and regulations around data while fostering an environment conducive to innovation and sustainability.
- **Balancing Innovation and Regulation**: The challenge will be to balance the need for **innovation** in agricultural technologies with the rights of farmers over their data. Policymakers will need to strike a balance between

encouraging technological advancements and preventing exploitation or monopolization of farmers' data by private companies.

- **Public-Private Partnerships**: Governments can play a significant role in fostering collaborative models for data governance. Public-private partnerships (PPPs) could facilitate the development of data-sharing platforms and the creation of ethical guidelines for data usage, ensuring that all stakeholders—farmers, tech companies, and consumers—benefit from a shared data ecosystem.

11.5 Data Security, Privacy, and the Digital Divide

- **Cybersecurity Threats**: As agricultural data becomes more valuable, cybersecurity will be a growing concern. Farmers and agricultural companies must adopt robust security measures to protect against data breaches, hacking, and unauthorized access. The challenge will be ensuring that smallholder farmers, who may not have access to advanced security technologies, can still protect their data effectively.
- **Privacy Concerns**: The growing use of personal data in agricultural technology raises serious privacy issues. Farmers' personal information, such as financial details, location data, and farm performance data, must be protected from misuse. As agricultural technology platforms grow, it will be important to implement strict privacy protocols and provide farmers with clear consent mechanisms.
- **Digital Divide**: While technological advances hold great promise, the digital divide remains a significant barrier. Farmers in rural or developing regions may not have the same access to technology, education, or infrastructure needed to participate in the data-driven agricultural economy. Addressing this divide is crucial for ensuring that data sovereignty is accessible to all farmers, regardless of their socio-economic status or geographical location.

11.6 The Future of Data-Driven Agriculture: Opportunities and Risks

- **Opportunities**: The future of data-driven agriculture offers immense opportunities for improving food security, sustainability, and economic development. Agricultural data can be used to optimize resource use, increase crop yields, reduce environmental impact, and promote more resilient farming practices. For farmers, this could translate into higher profits, lower costs, and better market access.
- **Risks**: However, there are significant risks associated with the growing reliance on agricultural data. These include the risk of data monopolies, unequal access, and the potential for data exploitation. If data sovereignty is not properly managed, smallholder farmers may find themselves at a disadvantage, with little control over how their data is used or monetized.

11.7 Recommendations for the Future

- **Establish Clear Legal Frameworks**: Governments and international bodies should work together to create clear and consistent legal frameworks that govern agricultural data sovereignty, focusing on farmers' rights, data protection, and fair compensation models.
- **Invest in Digital Literacy and Infrastructure**: To address the digital divide, governments and NGOs should invest in digital literacy programs and infrastructure development, ensuring that farmers in rural and underserved regions have access to the technologies and skills they need to participate in the data-driven agricultural economy.
- **Promote Collaborative Data Governance**: Encourage the development of data cooperatives, public-private partnerships, and farmer-led data initiatives to ensure that data governance remains transparent, ethical, and equitable for all stakeholders.

Summary

Chapter 11 focuses on the emerging trends and challenges that will shape the future of agricultural data sovereignty. With advancements in AI, big data, IoT, and precision agriculture, the agricultural sector is on the cusp of profound transformation. However, these opportunities come with significant ethical, regulatory, and security challenges. By proactively addressing these issues and fostering collaborative and innovative data governance models, stakeholders can ensure that the future of agricultural data remains inclusive, sustainable, and farmer-centric.

Chapter 12: Conclusion: The Road Ahead for Agricultural Data Sovereignty

Overview

In this final chapter, we summarize the key themes, challenges, and opportunities discussed throughout the book. We reflect on the importance of data sovereignty in shaping the future of agriculture, emphasizing the crucial role that farmers, policymakers, and industry players must play in ensuring that agricultural data is handled ethically, equitably, and sustainably. Looking ahead, this chapter provides a roadmap for the future, offering recommendations for fostering a data-driven agricultural landscape that respects farmers' rights and promotes global food security.

12.1 Reaffirming the Importance of Agricultural Data Sovereignty

Agricultural data sovereignty is not merely a regulatory issue but a fundamental pillar of agricultural sustainability and food security. As data continues to drive decision-making across the agricultural value chain, ensuring that farmers retain control over their data will be critical to fostering fair competition, economic empowerment, and environmental stewardship. The stakes are high: without the right to control and benefit from their data, farmers risk losing not just their economic agency but also their cultural heritage tied to agricultural practices.

- Farmers' rights to control their data are not just about individual autonomy; they are about ensuring that agricultural practices remain in the hands of those who depend on them the most.
- Local knowledge embedded in agricultural data needs protection and must be preserved within the context of indigenous and sustainable farming practices.

12.2 Key Challenges Revisited

Throughout the book, several significant challenges to agricultural data sovereignty have been identified:

- **Data Ownership and Control**: As technology giants and ag-tech companies gain increasing access to agricultural data, questions around who owns the data, how it is used, and how farmers can protect their interests remain unresolved.
- **Data Privacy and Security**: Ensuring that agricultural data is stored and used in a secure, privacy-respecting manner is vital to preserving farmer trust and encouraging broad participation in data-driven agricultural practices.
- **Regulatory Fragmentation**: The global nature of agricultural data flows, paired with inconsistent regulations, complicates the issue of cross-border data sharing. This leaves farmers and policymakers navigating a fragmented regulatory landscape.
- **Technological Barriers**: For many smallholder farmers, access to technology remains a significant barrier. Digital illiteracy, lack of infrastructure, and limited access to advanced data tools mean that the potential benefits of agricultural data are not always equitably distributed.

Despite these challenges, there are significant opportunities to address them through innovative policy frameworks, collaborative efforts, and technology-driven solutions.

12.3 Strategic Pathways Forward

To navigate the evolving landscape of agricultural data sovereignty, several strategic pathways need to be pursued:

- **Develop Comprehensive Data Sovereignty Frameworks**: Governments and international bodies should prioritize the development of comprehensive legal frameworks that define data ownership, usage rights, and fair compensation models. These frameworks should be flexible enough to adapt to emerging technologies while firmly protecting the rights of farmers.

- **Invest in Farmer-Led Data Models**: Encouraging farmer-driven data models, where farmers take the lead in managing their data, will be crucial. The use of blockchain and other decentralized technologies can help farmers protect their data and gain greater economic control over it.
- **Enhance Digital Literacy and Infrastructure**: Ensuring that rural farmers have access to the necessary digital tools and skills is vital for inclusive agricultural development. Public-private partnerships, government initiatives, and NGOs should work together to bridge the digital divide by providing affordable technology and training programs.
- **Promote Public-Private Collaborations**: Governments and industry players need to work collaboratively to design data-sharing platforms that empower farmers while fostering innovation. Public-private partnerships can help ensure fair access to agricultural data, encourage ethics-driven policies, and balance stakeholder interests.

12.4 Building a Future for All Stakeholders

The future of agricultural data sovereignty is not just about managing technology but about ensuring that all stakeholders—especially farmers—can benefit from data-driven practices in a way that is equitable and sustainable.

- **For Farmers**: Farmers need to be at the center of agricultural data governance. Ensuring that they have access to tools that allow them to control their data and share it selectively is crucial to ensuring their long-term viability. Farmers must have the ability to monetize their data while retaining full ownership of it.
- **For Policymakers**: Policymakers must create an inclusive regulatory framework that ensures data sovereignty while promoting innovation. By

developing multi-stakeholder platforms for dialogue, governments can ensure that all voices are heard and that regulatory approaches reflect the needs of the people.

- **For Industry**: Ag-tech companies and data-driven agricultural platforms should prioritize ethical considerations in their operations, adopting data-sharing models that align with the interests of farmers and other stakeholders. Moreover, they should work with governments to ensure that data is not monopolized, and farmers' rights are preserved.

12.5 Recommendations for Moving Forward

1. **Adopt Farmer-Centric Models**: All stakeholders should work towards creating models of data governance that are centered around the rights and needs of farmers. This includes developing farmer cooperatives for data management and enabling farmers to easily access their data for decision-making.
2. **Strengthen Global and Local Data Governance**: International and local stakeholders need to collaborate on establishing global agreements that ensure data so vereignty while allowing for responsible international data sharing. National policies should also be harmonized to enable efficient cross-border data flows while protecting farmers' rights.
3. **Foster Cross-Sector Partnerships**: Collaboration across sectors is essential to create robust data-sharing platforms that reflect ethical standards and meet the diverse needs of agricultural stakeholders. Industry players, governments, and farmers should form multi-stakeholder partnerships to drive data governance initiatives.
4. **Invest in Infrastructure and Digital Literacy**: Governments and organizations should prioritize investments in digital infrastructure and literacy programs to ensure that even the most marginalized farmers can take advantage of data-driven agricultural technologies.

12.6 Conclusion: Towards a Sustainable and Equitable Agricultural Future

Agricultural data sovereignty is not simply a matter of legal compliance but a critical foundation for building a sustainable and equitable agricultural future. As technology reshapes farming and food production, it is essential that farmers remain empowered, data rights are respected, and ethical standards guide the use of agricultural data. Through collaborative efforts, innovative policies, and farmer-centric models, we can ensure that agricultural data sovereignty remains a tool for global food security, environmental sustainability, and economic resilience.

Summary

Chapter 12 concludes the book by summarizing the key challenges and strategic solutions to advancing agricultural data sovereignty. Emphasizing the importance of farmer empowerment, inclusive regulation, and collaborative partnerships, it offers a vision for the future where farmers control their data, and data sovereignty drives ethical, sustainable, and equitable agricultural practices. The road ahead is challenging but filled with immense potential for reshaping agriculture into a more resilient and inclusive sector.

References

Bronson, K., & Knezevic, I. (2016). Big data in food and agriculture. *Big Data & Society, 3*(1), 1–5. https://doi.org/10.1177/2053951716648174

Carbonell, I. M. (2016). The ethics of big data in agriculture. *Internet Policy Review, 5*(1), 1–13. https://doi.org/10.14763/2016.1.405

Coble, K. H., Mishra, A. K., Ferrell, S., & Griffin, T. (2018). Big data in agriculture: A challenge for the future. *Applied Economic Perspectives and Policy, 40*(1), 79–96. https://doi.org/10.1093/aepp/ppx056

Eastwood, C., Klerkx, L., Ayre, M., & Dela Rue, B. (2019). Managing socio-ethical challenges in the development of smart farming: From a fragmented to a comprehensive approach for responsible research and innovation. *Journal of Agricultural and Environmental Ethics, 32*, 741–768. https://doi.org/10.1007/s10806-017-9704-5

European Commission. (2020). *A Farm to Fork Strategy for a fair, healthy and environmentally-friendly food system.* https://eur-lex.europa.eu/legal-content/EN/TXT/?uri=CELEX:52020DC0381

Ferris, H. (2020). Data sovereignty and agriculture. *Canadian Journal of Law and Technology, 18*(1), 1–28.

Food and Agriculture Organization of the United Nations (FAO). (2021). *Ethical principles for farm data governance*. https://www.fao.org/documents/card/en/c/CB3253EN/

HLPE (High Level Panel of Experts on Food Security and Nutrition). (2019). *Data collection and analysis tools for food security and nutrition*. Committee on World Food Security. https://www.fao.org/3/ca2587en/CA2587EN.pdf

Rotz, S., Gravely, E., Mosby, I., Duncan, E., Finnis, E., Horgan, M., & Fraser, E. D. G. (2019). Automated pastures and the digital divide: How agricultural technologies are shaping labour and rural communities. *Journal of Rural Studies, 68*, 112–122. https://doi.org/10.1016/j.jrurstud.2019.01.023

Shepherd, M., Turner, J. A., Small, B., & Wheeler, D. (2020). Priorities for science to overcome hurdles thwarting the full promise of the 'digital agriculture' revolution. *Journal of the Science of Food and Agriculture, 100*(14), 5083–5092. https://doi.org/10.1002/jsfa.9346

van der Burg, S., Bogaardt, M. J., & Wolfert, S. (2019). Ethics of smart farming: Current questions and directions for responsible innovation towards the future. *NJAS - Wageningen Journal of Life Sciences, 90–91*, 100289. https://doi.org/10.1016/j.njas.2019.01.001

Wiseman, L., Sanderson, J., Zhang, A., & Jakku, E. (2019). Farmers and their data: An examination of farmers' data rights in Australia. *Farm Policy Journal, 16*(1), 1–12.

Wolfert, S., Ge, L., Verdouw, C., & Bogaardt, M. J. (2017). Big data in smart farming–A review. *Agricultural Systems, 153*, 69–80. https://doi.org/10.1016/j.agsy.2017.01.023

Kolibri 360 GmbH
Ferdinand-Jühlke-Straße 7
99095 Erfurt, Deutschland
produktsicherheit@kolibri360.de

Druck:
CPI Druckdienstleistungen GmbH
im Auftrag der
Kolibri 360 GmbH
Ferdinand-Jühlke-Str. 7
99095 Erfurt